Lewis Carroll

The game of logic

Lewis Carroll

The game of logic

ISBN/EAN: 9783744738002

Printed in Europe, USA, Canada, Australia, Japan

Cover: Foto ©berggeist007 / pixelio.de

More available books at **www.hansebooks.com**

THE GAME

OF

LOGIC

BY

LEWIS CARROLL

PRICE THREE SHILLINGS

London
MACMILLAN AND CO.
AND NEW YORK
1887

To my Child-Friend.

I charm in vain; for never again,
All keenly as my glance I bend,
    Will Memory, goddess coy,
    Embody for my joy
Departed days, nor let me gaze
    On thee, my Fairy Friend!

Yet could thy face, in mystic grace,
A moment smile on me, 'twould send
    Far-darting rays of light
    From Heaven athwart the night,
By which to read in very deed
    Thy spirit, sweetest Friend!

So may the stream of Life's long dream
Flow gently onward to its end,
    With many a floweret gay,
    Adown its willowy way:
May no sigh vex, no care perplex,
    My loving little Friend!

NOTA BENE.

With each copy of this Book is given an Envelope, containing a Diagram (similar to the frontispiece) on card, and nine Counters, four red and five grey.

The Envelope, &c. can be had separately, at 3*d*. each.

The Author will be very grateful for suggestions, especially from beginners in Logic, of any alterations, or further explanations, that may seem desirable. Letters should be addressed to him at "29, Bedford Street, Covent Garden, London."

PREFACE

"There foam'd rebellious Logic, gagg'd and bound."

THIS Game requires nine Counters — four of one colour and five of another: say four red and five grey.

Besides the nine Counters, it also requires one Player, *at least*. I am not aware of any Game that can be played with *less* than this number: while there are several that require *more:* take Cricket, for instance, which requires twenty-two. How much easier it is, when you want to play a Game, to find *one* Player than twenty-two. At the same time, though one Player is enough, a good deal more amusement may be got by two working at it together, and correcting each other's mistakes.

A second advantage, possessed by this Game, is that, besides being an endless source of amusement (the number of arguments, that may be worked by it, being infinite), it will give the Players a little instruction as well. But is there any great harm in *that*, so long as you get plenty of amusement?

CONTENTS.

Chapter		Page
I.	**NEW LAMPS FOR OLD.**	
	§ 1. *Propositions*	1
	§ 2. *Syllogisms*.	20
	§ 3. *Fallacies*	32
II.	**CROSS QUESTIONS.**	
	§ 1. *Elementary*.	37
	§ 2. *Half of Smaller Diagram. Propositions to be represented*	40
	§ 3. *Do. Symbols to be interpreted* . . .	42
	§ 4. *Smaller Diagram. Propositions to be represented*	44
	§ 5. *Do. Symbols to be interpreted* . . .	46
	§ 6. *Larger Diagram. Propositions to be represented*	48
	§ 7. *Both Diagrams to be employed* . . .	51
III.	**CROOKED ANSWERS.**	
	§ 1. *Elementary*.	55
	§ 2. *Half of Smaller Diagram. Propositions represented*	59
	§ 3. *Do. Symbols interpreted*	61
	§ 4. *Smaller Diagram. Propositions represented.*	62
	§ 5. *Do. Symbols interpreted*	65
	§ 6. *Larger Diagram. Propositions represented.*	67
	§ 7. *Both Diagrams employed*	72
IV.	**HIT OR MISS**	85

CHAPTER I.

NEW LAMPS FOR OLD.

"Light come, light go."

§ 1. *Propositions.*

"Some new Cakes are nice."
"No new Cakes are nice."
"All new Cakes are nice."

There are three '*Propositions*' for you——the only three kinds we are going to use in this Game: and the first thing to be done is to learn how to express them on the Board.

Let us begin with
"Some new Cakes are nice."

But, before doing so, a remark has to be made——one that is rather important, and by no means easy to understand all in a moment: so please to read this *very* carefully.

The world contains many *Things* (such as "Buns", "Babies", "Beetles", "Battledores", &c.); and the Things possess many *Attributes* (such as "baked", "beautiful", "black", "broken", &c.: in fact, whatever can be "attributed to", that is "said to belong to", any Thing, is an Attribute). Whenever we wish to mention a Thing, we use a *Substantive*: when we wish to mention an Attribute, we use an *Adjective*. People have asked the question "Can a Thing exist without any Attributes belonging to it?" It is a very puzzling question, and I'm not going to try to answer it: let us turn up our noses, and treat it with contemptuous silence, as if it really wasn't worth noticing. But, if they put it the other way, and ask "Can an Attribute exist without any Thing for it to belong to?", we may say at once "No: no more than a Baby could go a railway-journey with no one to take care of it!" You never saw "beautiful" floating about in the air, or littered about on the floor, without any Thing to *be* beautiful, now did you?

And now what am I driving at, in all this long rigmarole? It is this. You may put "is" or "are" between the names of two *Things* (for example, "some Pigs are fat Animals"), or between the names of two *Attributes* (for example, "pink is light-red"), and in each case it will make good sense. But, if you put "is" or "are" between the name of a *Thing* and the name of an *Attribute* (for example, "some Pigs are pink "), you do *not* make good sense (for how can a Thing *be* an Attribute?) unless you have an understanding with the person to whom you are speaking. And the simplest understanding would, I think, be this——that the Substantive shall be supposed to be repeated at the end of the sentence, so that the sentence, if written out in full, would be "some Pigs are pink (Pigs)". And now the word "are" makes quite good sense.

Thus, in order to make good sense of the Proposition "some new Cakes are nice", we must suppose it to be written out in full, in the form "some new Cakes are nice (Cakes)". Now this contains two '*Terms*'—— "new Cakes" being one of them, and "nice (Cakes)" the other. "New Cakes," being the one we are talking about, is called the '*Subject*' of the Proposition, and "nice (Cakes)" the '*Predicate*'. Also this Proposition is said to be a '*Particular*' one, since it does not speak of the *whole* of its Subject, but only of a *part* of it. The other two kinds are said to be '*Universal*', because they speak of the *whole* of their Subjects——the one denying niceness, and the other asserting it, of the *whole* class of "new Cakes". Lastly, if you would like to have a definition of the word '*Proposition*' itself, you may take this:—"a sentence stating that some, or none, or all, of the Things belonging to a certain class, called its 'Subject', are also Things belonging to a certain other class, called its 'Predicate'".

You will find these seven words —— *Proposition, Attribute, Term, Subject, Predicate, Particular, Universal* ——charmingly useful, if any friend should happen to ask if you have ever studied Logic. Mind you bring all seven words into your answer, and your friend will go away deeply impressed——'a sadder and a wiser man'.

Now please to look at the smaller Diagram on the Board, and suppose it to be a cupboard, intended for all the Cakes in the world (it would have to be a good large one, of course). And let us suppose all the new ones to be put into the upper half (marked 'x'), and all the rest (that is, the *not*-new ones) into the lower half (marked 'x''). Thus the lower half would contain *elderly* Cakes, *aged* Cakes, *ante-diluvian* Cakes——if there are any: I haven't seen many, myself——and so on. Let us also suppose all the nice Cakes to be put into the left-hand half (marked 'y'), and all the rest (that is, the not-nice ones) into the right-hand half (marked 'y''). At present, then, we must understand x to mean "new", x' "not-new", y "nice", and y' "not-nice."

And now what kind of Cakes would you expect to find in compartment No. 5?

It is part of the upper half, you see; so that, if it has any Cakes in it, they must be *new*: and it is part of the left-hand half; so that they must be *nice*. Hence if there are any Cakes in this compartment, they must have the double '*Attribute*' "new and nice": or, if we use letters, they must be "$x\, y$."

Observe that the letters x, y are written on two of the edges of this compartment. This you will find a very convenient rule for knowing what Attributes belong to the Things in any compartment. Take No. 7, for instance. If there are any Cakes there, they must be "$x'\, y$", that is, they must be "not-new and nice."

Now let us make another agreement——that a red counter in a compartment shall mean that it is '*occupied*', that is, that there are *some* Cakes in it. (The word 'some,' in Logic, means 'one or more': so that a single Cake in a compartment would be quite enough reason for saying "there are *some* Cakes here"). Also let us agree that a grey counter in a compartment shall mean that it is '*empty*', that is, that there are *no* Cakes in it. In the following Diagrams, I shall put '1' (meaning 'one or more') where you are to put a *red* counter, and '0' (meaning 'none') where you are to put a *grey* one.

As the Subject of our Proposition is to be "new Cakes", we are only concerned, at present, with the *upper* half of the cupboard, where all the Cakes have the attribute x, that is, "new."

Now, fixing our attention on this upper half, suppose we found it marked like this,

1	

that is, with a red counter in No. 5. What would this tell us, with regard to the class of " new Cakes " ?

Would it not tell us that there are *some* of them in the x y-compartment? That is, that some of them (besides having the Attribute x, which belongs to both compartments) have the Attribute y (that is, "nice"). This we might express by saying " some x-Cakes are y-(Cakes) ", or, putting words instead of letters,

"Some new Cakes are nice (Cakes)",

or, in a shorter form,

" Some new Cakes are nice ".

At last we have found out how to represent the first Proposition of this Section. If you have not *clearly* understood all I have said, go no further, but read it over and over again, till you *do* understand it. After that is once mastered, you will find all the rest quite easy.

It will save a little trouble, in doing the other Propositions, if we agree to leave out the word "Cakes" altogether. I find it convenient to call the whole class of Things, for which the cupboard is intended, the ' *Universe.*' Thus we might have begun this business by saying "Let us take a Universe of Cakes." (Sounds nice, doesn't it ?)

Of course any other Things would have done just as well as Cakes. We might make Propositions about "a Universe of Lizards", or even "a Universe of Hornets". (Wouldn't *that* be a charming Universe to live in?)

So far, then, we have learned that

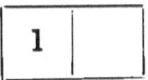

means "some x and y," i. e. "some new are nice."

I think you will see, without further explanation, that

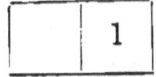

means "some x are y'," i. e. "some new are not-nice."

Now let us put a *grey* counter into No. 5, and ask ourselves the meaning of

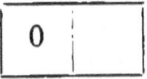

This tells us that the $x\,y$-compartment is *empty*, which we may express by "no x are y", or, "no new Cakes are nice". This is the second of the three Propositions at the head of this Section.

In the same way,

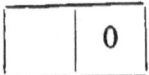

would mean "no x are y'," or, "no new Cakes are not-nice."

What would you make of this, I wonder?

$$\boxed{1 \mid 1}$$

I hope you will not have much trouble in making out that this represents a *double* Proposition: namely, "some x are y, *and* some are y'," i. e. "some new are nice, *and* some are not-nice."

The following is a little harder, perhaps:—

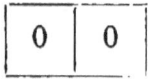

This means "no x are y, *and* none are y'," i. e. "no new are nice, *and* none are not-nice": which leads to the rather curious result that "no new exist," i.e. "no Cakes are new." This is because "nice" and "not-nice" make what we call an '*exhaustive*' division of the class "new Cakes": i. e. between them, they *exhaust* the whole class, so that all the new Cakes, that exist, must be found in one or the other of them.

And now suppose you had to represent, with counters, the contradictory to "no Cakes are new", which would be "some Cakes are new", or, putting letters for words, "some Cakes are x", how would you do it?

This will puzzle you a little, I expect. Evidently you must put a red counter *somewhere* in the x-half of the cupboard, since you know there are *some* new Cakes. But you must not put it into the *left-hand* compartment, since you do not know them to be *nice*: nor may you put it into the *right-hand* one, since you do not know them to be *not-nice*.

What, then, are you to do? I think the best way out of the difficulty is to place the red counter *on the division-line* between the xy-compartment and the xy'-compartment. This I shall represent (as *I* always put '1' where *you* are to put a red counter) by the diagram

—	—

Our ingenious American cousins have invented a phrase to express the position of a man who wants to join one or other of two parties——such as their two parties 'Democrats' and 'Republicans'——but ca'n't make up his mind *which*. Such a man is said to be "sitting on the fence." Now that is exactly the position of the red counter you have just placed on the division-line. He likes the look of No. 5, and he likes the look of No. 6, and he doesn't know *which* to jump down into. So there he sits astride, silly fellow, dangling his legs, one on each side of the fence!

Now I am going to give you a much harder one to make out. What does this mean?

1	0

This is clearly a *double* Proposition. It tells us, not only that "some x are y," but also that "no x are *not* y." Hence the result is "*all* x are y," i. e. "all new Cakes are nice", which is the last of the three Propositions at the head of this Section.

We see, then, that the Universal Proposition
"All new Cakes are nice"
consists of *two* Propositions taken together, namely,
"Some new Cakes are nice,"
and "No new Cakes are not-nice."

In the same way

| 0 | 1 |

would mean "all x are y'", that is,
"All new Cakes are not-nice."

Now what would you make of such a Proposition as "The Cake you have given me is nice"? Is it Particular, or Universal?

"Particular, of course," you readily reply. "One single Cake is hardly worth calling 'some,' even."

No, my dear impulsive Reader, it is 'Universal'. Remember that, few as they are (and I grant you they couldn't well be fewer), they are (or rather 'it is') *all* that you have given me! Thus, if (leaving 'red' out of the question) I divide my Universe of Cakes into two classes——the Cakes you have given me (to which I assign the upper half of the cupboard), and those you *haven't* given me (which are to go below)—— I find the lower half fairly full, and the upper one as nearly as possible empty. And then, when I am told to put an upright division into each half, keeping the *nice* Cakes to the left, and the *not-nice* ones to the right, I begin by carefully collecting *all* the Cakes you have given me (saying to myself, from time to time, "Generous creature! How shall I ever repay such kindness?"), and piling them up in the left-hand compartment. *And it doesn't take long to do it!*

Here is another Universal Proposition for you. "Barzillai Beckalegg is an honest man." That means "*All* the Barzillai Beckaleggs, that I am now considering, are honest men." (You think I invented that name, now don't you? But I didn't. It's on a carrier's cart, somewhere down in Cornwall.)

This kind of Universal Proposition (where the Subject is a single Thing) is called an '*Individual*' Proposition.

Now let us take "*nice* Cakes" as the Subject of our Proposition: that is, let us fix our thoughts on the *left-hand* half of the cupboard, where all the Cakes have the attribute y, that is, "nice."
Suppose we find it marked like this:—
What would that tell us?

I hope that it is not necessary, after explaining the *horizontal* oblong so fully, to spend much time over the *upright* one. I hope you will see, for yourself, that this means " some y are x ", that is,
"Some nice Cakes are new."

"But," you will say, "we have had this case before. You put a red counter into No. 5, and you told us it meant 'some new Cakes are nice'; and *now* you tell us that it means 'some *nice* Cakes are *new*'! Can it mean *both*?"

The question is a very thoughtful one, and does you *great* credit, dear Reader! It *does* mean both. If you choose to take x (that is, "new Cakes") as your Subject, and to regard No. 5 as part of a *horizontal* oblong, you may read it "some x are y", that is, "some new Cakes are nice" : but, if you choose to take y (that is, "nice Cakes") as your Subject, and to regard No. 5 as part of an *upright* oblong, *then* you may read it "some y are x", that is, "some nice Cakes are new". They are merely two different ways of expressing the very same truth.

Without more words, I will simply set down the other ways in which this upright oblong might be marked, adding the meaning in each case. By comparing them with the various cases of the horizontal oblong, you will, I hope, be able to understand them clearly.

You will find it a good plan to examine yourself on this table, by covering up first one column and then the other, and 'dodging about', as the children say.

Also you will do well to write out for yourself two other tables——one for the *lower* half of the cupboard, and the other for its *right-hand* half.

And now I think we have said all we need to say about the smaller Diagram, and may go on to the larger one.

Symbols.	Meanings.
 ― 1	Some y are x'; i.e. Some nice are not-new.
0 ― 	No y are x; i.e. No nice are new. [Observe that this is merely another way of expressing " No new are nice."]
 ― 0	No y are x'; i.e. No nice are not-new.
1 ― 1	Some y are x, and some are x'; i.e. Some nice are new, and some are not-new.
0 ― 0	No y are x, and none are x'; i.e. No y exist; i.e. No Cakes are nice.
1 ― 0	All y are x; i.e. All nice are new.
0 ― 1	All y are x'; i.e. All nice are not-new.

This may be taken to be a cupboard divided in the same way as the last, but *also* divided into two portions, for the Attribute m. Let us give to m the meaning "wholesome": and let us suppose that all *wholesome* Cakes are placed *inside* the central Square, and all the *unwholesome* ones *outside* it, that is, in one or other of the four queer-shaped *outer* compartments.

We see that, just as, in the smaller Diagram, the Cakes in each compartment had *two* Attributes, so, here, the Cakes in each compartment have *three* Attributes: and, just as the letters, representing the *two* Attributes, were written on the *edges* of the compartment, so, here, they are written at the *corners*. (Observe that m' is supposed to be written at each of the four outer corners.) So that we can tell in a moment, by looking at a compartment, what three Attributes belong to the Things in it. For instance, take No. 12. Here we find x, y', m, at the corners: so we know that the Cakes in it, if there are any, have the triple Attribute, '$xy'm$', that is, "new, not-nice, and wholesome." Again, take No. 16. Here we find, at the corners, x', y', m': so the Cakes in it are "not-new, not-nice, and unwholesome." (Remarkably untempting Cakes!)

It would take far too long to go through all the Propositions, containing x and y, x and m, and y and m, which can be represented on this diagram (there are ninety-six altogether, so I am sure you will excuse me!)

§ 1.]   PROPOSITIONS.   15

and I must content myself with doing two or three, as specimens. You will do well to work out a lot more for yourself.

Taking the upper half by itself, so that our Subject is " new Cakes ", how are we to represent " no new Cakes are wholesome " ?

This is, writing letters for words, " no x are m." Now this tells us that none of the Cakes, belonging to the upper half of the cupboard, are to be found *inside* the central Square: that is, the two compartments, No. 11 and No. 12, are *empty*. And this, of course, is represented by

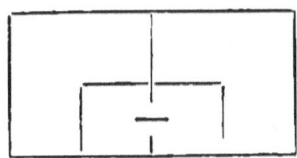

And now how are we to represent the contradictory Proposition "*some* x are m" ? This is a difficulty I have already considered. I think the best way is to place a red counter *on the division-line* between No. 11 and No. 12, and to understand this to mean that *one* of the two compartments is 'occupied,' but that we do not at present know *which*. This I shall represent thus :—

Now let us express "all x are m."

This consists, we know, of *two* Propositions,
"Some x are m,"
and "No x are m'."

Let us express the negative part first. This tells us that none of the Cakes, belonging to the upper half of the cupboard, are to be found *outside* the central Square: that is, the two compartments, No. 9 and No. 10, are *empty*. This, of course, is represented by

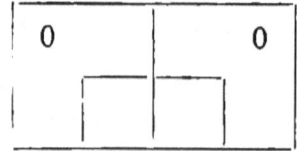

But we have yet to represent "Some x are m." This tells us that there are *some* Cakes in the oblong consisting of No. 11 and No. 12: so we place our red counter, as in the previous example, on the division-line between No. 11 and No. 12, and the result is

Now let us try one or two interpretations.

What are we to make of this, with regard to x and y?

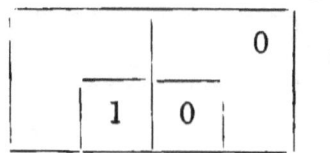

§ 1.]   PROPOSITIONS.   17

This tells us, with regard to the xy'-Square, that it is wholly 'empty', since *both* compartments are so marked. With regard to the xy-Square, it tells us that it is 'occupied'. True, it is only *one* compartment of it that is so marked; but that is quite enough, whether the other be 'occupied' or 'empty', to settle the fact that there is *something* in the Square.

If, then, we transfer our marks to the smaller Diagram, so as to get rid of the m-subdivisions, we have a right to mark it

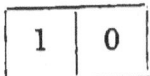

which means, you know, "all x are y."

The result would have been exactly the same, if the given oblong had been marked thus :—

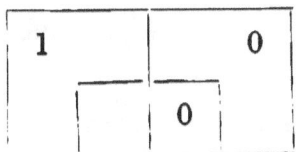

Once more: how shall we interpret this, with regard to x and y?

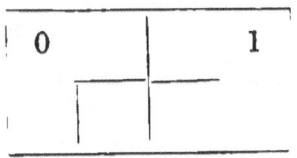

This tells us, as to the xy-Square, that *one* of its compartments is 'empty'. But this information is

c

quite useless, as there is no mark in the *other* compartment. If the other compartment happened to be 'empty' too, the Square would be 'empty': and, if it happened to be 'occupied', the Square would be 'occupied'. So, as we do not know *which* is the case, we can say nothing about *this* Square.

The other Square, the xy'-Square, we know (as in the previous example) to be 'occupied'.

If, then, we transfer our marks to the smaller Diagram, we get merely this :—

	1

which means, you know, "some x are y'."

These principles may be applied to all the other oblongs. For instance, to represent "all y' are m'" we should mark the *right-hand upright oblong* (the one that has the attribute y') thus:—

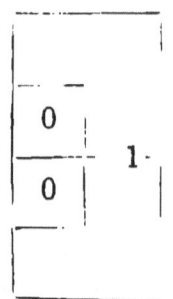

and, if we were told to interpret the lower half of the cupboard, marked as follows, with regard to x and y,

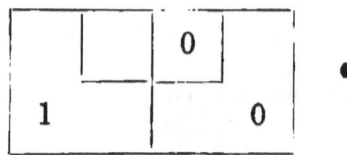

we should transfer it to the smaller Diagram thus,

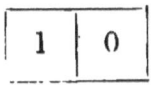

and read it "all x' are y."

Two more remarks about Propositions need to be made.

One is that, in every Proposition beginning with "some" or "all", the *actual existence* of the 'Subject' is asserted. If, for instance, I say "all misers are selfish," I mean that misers *actually exist*. If I wished to avoid making this assertion, and merely to state the *law* that miserliness necessarily involves selfishness, I should say "no misers are unselfish" which does not assert that any misers exist at all, but merely that, if any *did* exist, they *would* be selfish.

The other is that, when a Proposition begins with "some" or "no", and contains more than two Attributes, these Attributes may be re-arranged, and shifted from one Term to the other, *ad libitum*. For example, "some abc are def" may be re-arranged as "some bf are $acde$," each being equivalent to "some Things are $abcdef$". Again "No wise old men are rash and reckless gamblers" may be re-arranged as "No rash old gamblers are wise and reckless," each being equivalent to "No men are wise old rash reckless gamblers."

§ 2. *Syllogisms.*

Now suppose we divide our Universe of Things in three ways, with regard to three different Attributes. Out of these three Attributes, we may make up three different couples (for instance, if they were *a, b, c*, we might make up the three couples *ab, ac, bc*). Also suppose we have two Propositions given us, containing two of these three couples, and that from them we can prove a third Proposition containing the third couple. (For example, if we divide our Universe for *m, x,* and *y*; and if we have the two Propositions given us, "no *m* are *x'*" and "all *m'* are *y*", containing the two couples *mx* and *my*, it might be possible to prove from them a third Proposition, containing *x* and *y*.)

In such a case we call the given Propositions '*the Premisses*', the third one '*the Conclusion*' and the whole set '*a Syllogism*'.

Evidently, *one* of the Attributes must occur in both Premisses; or else one must occur in *one* Premiss, and its *contradictory* in the other.

In the first case (when, for example, the Premisses are "some m are x" and "no m are y'") the Term, which occurs twice, is called '*the Middle Term*', because it serves as a sort of link between the other two Terms.

In the second case (when, for example, the Premisses are "no m are x'" and "all m' are y") the two Terms, which contain these contradictory Attributes, may be called '*the Middle Terms*'.

Thus, in the first case, the class of "m-Things" is the Middle Term; and, in the second case, the two classes of "m-Things" and "m'-Things" are the Middle Terms.

The Attribute, which occurs in the Middle Term or Terms, disappears in the Conclusion, and is said to be "eliminated", which literally means "turned out of doors".

Now let us try to draw a Conclusion from the two Premisses—

"Some new Cakes are unwholesome;
No nice Cakes are unwholesome."

In order to express them with counters, we need to divide Cakes in *three* different ways, with regard to newness, to niceness, and to wholesomeness. For this we must use the larger Diagram, making x mean "new", y "nice", and m "wholesome". (Everything *inside* the central Square is supposed to have the attribute *m*, and everything *outside* it the attribute *m'*, i.e. "not-*m*".)

You had better adopt the rule to make *m* mean the Attribute which occurs in the *Middle* Term or Terms. (I have chosen *m* as the symbol, because 'middle' begins with 'm'.)

Now, in representing the two Premisses, I prefer to begin with the *negative* one (the one beginning with "no"), because *grey* counters can always be placed with *certainty*, and will then help to fix the position of the red counters, which are sometimes a little uncertain where they will be most welcome.

Let us express, then, "no nice Cakes are unwholesome (Cakes)", i.e. "no *y*-Cakes are *m'*-(Cakes)". This tells us that none of the Cakes belonging to the *y*-half of the cupboard are in its *m'*-compartments (i.e. the ones *outside* the central Square). Hence the two compartments, No. 9 and No. 15, are both '*empty*'; and we must place a grey counter in *each* of them, thus :—

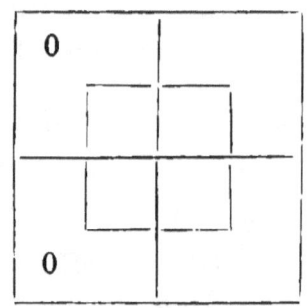

[§ 2.] SYLLOGISMS. 23

We have now to express the other Premiss, namely, "some new Cakes are unwholesome (Cakes)", i.e. "some x-Cakes are m'-(Cakes)". This tells us that some of the Cakes in the x-half of the cupboard are in its m'-compartments. Hence *one* of the two compartments, No. 9 and No. 10, is 'occupied': and, as we are not told in *which* of these two compartments to place the red counter, the usual rule would be to lay it on the division-line between them: but, in this case, the other Premiss has settled the matter for us, by declaring No. 9 to be *empty*. Hence the red counter has no choice, and *must* go into No. 10, thus:—

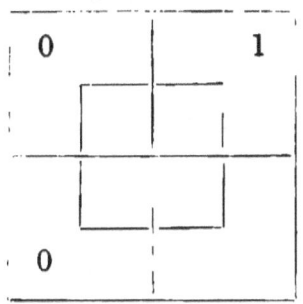

And now what counters will this information enable us to place in the *smaller* Diagram, so as to get some Proposition involving x and y only, leaving out m? Let us take its four compartments, one by one.

First, No. 5. All we know about *this* is that its *outer* portion is empty: but we know nothing about its *inner* portion. Thus the Square *may* be empty, or it *may* have something in it. Who can tell? So we dare not place *any* counter in this Square.

Secondly, what of No. 6? Here we are a little better off. We know that there is *something* in it, for there is a red counter in its outer portion. It is true we do not know whether its inner portion is empty or occupied: but what does *that* matter? One solitary Cake, in one corner of the Square, is quite sufficient excuse for saying "*this Square is occupied*", and for marking it with a red counter.

As to No. 7, we are in the same condition as with No. 5——we find it *partly* 'empty', but we do not know whether the other part is empty or occupied: so we dare not mark this Square.

And as to No. 8, we have simply no information *at all.*

The result is

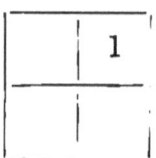

Our 'Conclusion', then, must be got out of the rather meagre piece of information that there is a red counter in the xy'-Square. Hence our Conclusion is "some x are y'", i.e. "some new Cakes are not-nice (Cakes)": or, if you prefer to take y' as your Subject, "some not-nice Cakes are new (Cakes)"; but the other looks neatest.

We will now write out the whole Syllogism, putting the symbol ∴ for "therefore", and omitting "Cakes", for the sake of brevity, at the end of each Proposition.

"Some new Cakes are unwholesome;
No nice Cakes are unwholesome.
∴ Some new Cakes are not-nice."

And you have now worked out, successfully, your first '*Syllogism*'. Permit me to congratulate you, and to express the hope that it is but the beginning of a long and glorious series of similar victories!

We will work out one other Syllogism——a rather harder one than the last——and then, I think, you may be safely left to play the Game by yourself, or (better) with any friend whom you can find, that is able and willing to take a share in the sport.

Let us see what we can make of the two Premisses—

"All Dragons are uncanny;
All Scotchmen are canny."

Remember, I don't guarantee the Premisses to be *facts*. In the first place, I never even saw a Dragon: and, in the second place, it isn't of the slightest consequence to us, as *Logicians*, whether our Premisses are true or false: all *we* have to do is to make out whether they *lead logically to the Conclusion*, so that, if *they* were true, *it* would be true also.

You see, we must give up the "Cakes" now, or our cupboards will be of no use to us. We must take, as our 'Universe', some class of things which will include Dragons and Scotchmen: shall we say 'Animals'? And, as "canny" is evidently the Attribute belonging to the 'Middle Terms', we will let *m* stand for "canny", *x* for "Dragons", and *y* for "Scotchmen". So that our two Premisses are, in full,

"All Dragon-Animals are uncanny (Animals);
All Scotchman-Animals are canny (Animals)."

And these may be expressed, using letters for words, thus :—

"All *x* are *m'*;
All *y* are *m*."

The first Premiss consists, as you already know, of two parts :—

"Some *x* are *m'*,"
and "No *x* are *m*."

And the second also consists of two parts :—

"Some *y* are *m*,"
and "No *y* are *m'*."

Let us take the negative portions first.

We have, then, to mark, on the larger Diagram, first, "no *x* are *m*", and secondly, "no *y* are *m'*". I think you will see, without further explanation, that the two results, separately, are

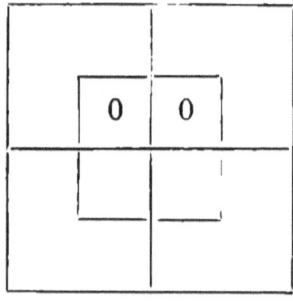 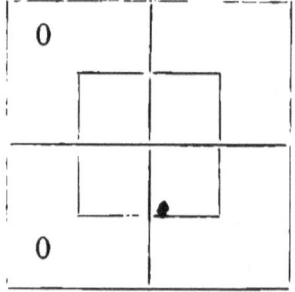

§ 2.]   *SYLLOGISMS.*   27

and that these two, when combined, give us

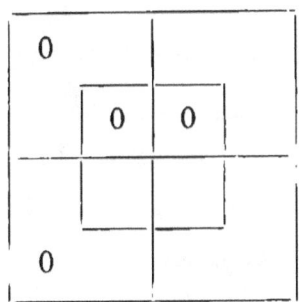

We have now to mark the two positive portions, "some x are m'" and "some y are m".

The only two compartments, available for Things which are xm', are No. 9 and No. 10. Of these, No. 9 is already marked as 'empty'; so our red counter *must* go into No. 10.

Similarly, the only two, available for ym, are No. 11 and No. 13. Of these, No. 11 is already marked as 'empty'; so our red counter *must* go into No. 13.

The final result is

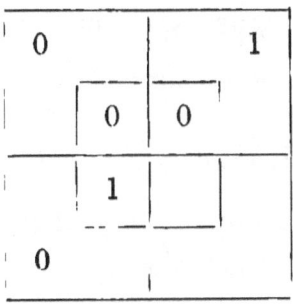

And now how much of this information can usefully be transferred to the smaller Diagram?

Let us take its four compartments, one by one.

As to No. 5? This, we see, is wholly 'empty'. (So mark it with a grey counter.)

As to No. 6? This, we see, is 'occupied'. (So mark it with a red counter.)

As to No. 7? Ditto, ditto.

As to No. 8? No information.

The smaller Diagram is now pretty liberally marked :—

0	1
1	

And now what Conclusion can we read off from this? Well, it is impossible to pack such abundant information into *one* Proposition: we shall have to indulge in *two*, this time.

First, by taking x as Subject, we get "all x are y'", that is,
"All Dragons are not-Scotchmen":
secondly, by taking y as Subject, we get "all y are x'", that is,
"All Scotchmen are not-Dragons".

Let us now write out, all together, our two Premisses and our brace of Conclusions.

> "All Dragons are uncanny;
> All Scotchmen are canny.
>
> ∴ { All Dragons are not-Scotchmen;
>   All Scotchmen are not-Dragons."

Let me mention, in conclusion, that you may perhaps meet with logical treatises in which it is not assumed that any Thing *exists* at all, but "some x are y" is understood to mean "the Attributes x, y are *compatible*, so that a Thing can have both at once", and "no x are y" to mean "the Attributes x, y are *incompatible*, so that nothing can have both at once".

In such treatises, Propositions have quite different meanings from what they have in our 'Game of Logic', and it will be well to understand exactly what the difference is.

First take "some x are y". Here *we* understand "are" to mean "are, as an actual *fact*"——which of course implies that some x-Things *exist*. But *they* (the writers of these other treatises) only understand "are" to mean "*can* be", which does not at all imply that any *exist*. So they mean *less* than we do: our meaning includes theirs (for of course "some x *are* y" includes "some x *can be* y"), but theirs does *not* include ours. For example, "some Welsh hippopotami are heavy" would be *true*, according to these writers (since the

Attributes "Welsh" and "heavy" are quite *compatible* in a hippopotamus), but it would be *false* in our Game (since there are no Welsh hippopotami to *be* heavy).

Secondly, take "no x are y". Here *we* only understand "are" to mean "are, as an actual *fact*"——which does not at all imply that no x *can* be y. But *they* understand the Proposition to mean, not only that none *are* y, but that none *can possibly* be y. So they mean *more* than we do: their meaning includes ours (for of course "no x *can* be y" includes "no x *are* y"), but ours does *not* include theirs. For example, "no Policemen are eight feet high" would be *true* in our Game (since, as an actual fact, no such splendid specimens are ever found), but it would be *false*, according to these writers (since the Attributes "belonging to the Police Force" and "eight feet high" are quite *compatible*: there is nothing to *prevent* a Policeman from growing to that height, if sufficiently rubbed with Rowland's Macassar Oil——which is said to make *hair* grow, when rubbed on hair, and so of course will make a *Policeman* grow, when rubbed on a Policeman).

Thirdly, take "all x are y", which consists of the two partial Propositions "some x are y" and "no x are y'". Here, of course, the treatises mean *less* than we do in the *first* part, and *more* than we do in the *second*. But the two operations don't balance each other——

[§ 2.] SYLLOGISMS.

any more than you can console a man, for having knocked down one of his chimneys, by giving him an extra door-step.

If you meet with Syllogisms of this kind, you may work them, quite easily, by the system I have given you: you have only to make 'are' mean 'are *capable of being*', and all will go smoothly. For "some x are y" will become "some x are capable of being y", that is, "the Attributes x, y are *compatible*". And "no x are y" will become "no x are capable of being y", that is, "the Attributes x, y are *incompatible*". And, of course, "all x are y" will become "some x are capable of being y, and none are capable of being y'", that is, "the Attributes x, y are *compatible*, and the Attributes x, y' are *incompatible*." In using the Diagrams for this system, you must understand a red counter to mean "there may *possibly* be something in this compartment," and a grey one to mean "there cannot *possibly* be anything in this compartment."

§ 3. *Fallacies.*

And so you think, do you, that the chief use of Logic, in real life, is to deduce Conclusions from workable Premisses, and to satisfy yourself that the Conclusions, deduced by other people, are correct? I only wish it were! Society would be much less liable to panics and other delusions, and *political* life, especially, would be a totally different thing, if even a majority of the arguments, that are scattered broadcast over the world, were correct! But it is all the other way, I fear. For *one* workable Pair of Premisses (I mean a Pair that lead to a logical Conclusion) that you meet with in reading your newspaper or magazine, you will probably find *five* that lead to no Conclusion at all: and, even when the Premisses *are* workable, for *one* instance, where the writer draws a correct Conclusion, there are probably *ten* where he draws an incorrect one.

In the first case, you may say "the *Premisses* are fallacious": in the second, "the *Conclusion* is fallacious."

The chief use you will find, in such Logical skill as this Game may teach you, will be in detecting '*Fallacies*' of these two kinds.

The first kind of Fallacy——'Fallacious Premisses'—— you will detect when, after marking them on the larger Diagram, you try to transfer the marks to the smaller. You will take its four compartments, one by one, and ask, for each in turn, "What mark can I place *here*?"; and in *every* one the answer will be "No information!", showing that there is *no Conclusion at all*. For instance,

"All soldiers are brave;
Some Englishmen are brave.
∴ Some Englishmen are soldiers."

looks uncommonly *like* a Syllogism, and might easily take in a less experienced Logician. But *you* are not to be caught by such a trick! You would simply set out the Premisses, and would then calmly remark "Fallacious *Premisses*!": you wouldn't condescend to ask what *Conclusion* the writer professed to draw——knowing that, *whatever* it is, it *must* be wrong. You would be just as safe as that wise mother was, who said "Mary, just go up to the nursery, and see what Baby's doing, *and tell him not to do it*!"

The other kind of Fallacy——'Fallacious Conclusion' ——you will not detect till you have marked *both* Diagrams, and have read off the correct Conclusion, and have compared it with the Conclusion which the writer has drawn.

But mind, you mustn't say "*Fallacious* Conclusion," simply because it is not *identical* with the correct one: it may be a *part* of the correct Conclusion, and so be quite correct, *as far as it goes*. In this case you would merely remark, with a pitying smile, "*Defective* Conclusion!" Suppose, for example, you were to meet with this Syllogism :—

"All unselfish people are generous;
No misers are generous.
∴ No misers are unselfish."

the Premisses of which might be thus expressed in letters :—

"All x' are m;
No y are m."

Here the correct Conclusion would be "All x' are y'" (that is, "All unselfish people are not misers"), while the Conclusion, drawn by the writer, is "No y are x'," (which is the same as "No x' are y," and so is *part* of "All x' are y'.") Here you would simply say "*Defective* Conclusion!" The same thing would happen, if you were in a confectioner's shop, and if a little boy were to come in, put down twopence, and march off triumphantly with a single penny-bun. You would shake your head mournfully, and would remark "Defective Conclusion! Poor little chap!" And perhaps you would ask the young lady behind the counter whether she would let *you* eat the bun, which the little boy had paid for and left behind him: and perhaps *she* would reply "Sha'n't!"

But if, in the above example, the writer had drawn the Conclusion "All misers are selfish" (that is, "All y are x"), this would be going *beyond* his legitimate rights (since it would assert the *existence* of y, which is not contained in the Premisses), and you would very properly say "Fallacious Conclusion!"

Now, when you read other treatises on Logic, you will meet with various kinds of (so-called) 'Fallacies', which are by no means *always* so. For example, if you were to put before one of these Logicians the Pair of Premisses

"No honest men cheat;
No dishonest men are trustworthy."

and were to ask him what Conclusion followed, he would probably say "None at all! Your Premisses offend against *two* distinct Rules, and are as fallacious as they can well be!" Then suppose you were bold enough to say "The Conclusion is 'No men who cheat are trustworthy'," I fear your Logical friend would turn away hastily——perhaps angry, perhaps only scornful: in any case, the result would be unpleasant. *I advise you not to try the experiment!*

"But why is this?" you will say. "Do you mean to tell us that all these Logicians are wrong?" Far from it, dear Reader! From *their* point of view, they are perfectly right. But they do not include, in their system, anything like *all* the possible forms of Syllogisms.

They have a sort of nervous dread of Attributes beginning with a negative particle. For example, such Propositions as "All not-x are y," "No x are not-y," are quite outside their system. And thus, having (from sheer nervousness) excluded a quantity of very useful forms, they have made rules which, though quite applicable to the few forms which they allow of, are no use at all when you consider all possible forms.

Let us not quarrel with them, dear Reader! There is room enough in the world for both of us. Let us quietly take our broader system: and, if they choose to shut their eyes to all these useful forms, and to say "They are not Syllogisms at all!" we can but stand aside, and let them Rush upon their Fate! There is scarcely anything of yours, upon which it is so dangerous to Rush, as your Fate. You may Rush upon your Potato-beds, or your Strawberry-beds, without doing much harm: you may even Rush upon your Balcony (unless it is a new house, built by contract, and with no clerk of the works) and may survive the foolhardy enterprise: but if you once Rush upon your *Fate*—why, you must take the consequences!

CHAPTER II.

CROSS QUESTIONS.

*"The Man in the Wilderness asked of me
'How many strawberries grow in the sea?'"*

---

§ 1. *Elementary.*

1. What is an 'Attribute'? Give examples.

2. When is it good sense to put "is" or "are" between two names? Give examples.

3. When is it *not* good sense? Give examples.

4. When it is *not* good sense, what is the simplest agreement to make, in order to make good sense?

5. Explain 'Proposition', 'Term', 'Subject', and 'Predicate'. Give examples.

6. What are 'Particular' and 'Universal' Propositions? Give examples.

7. Give a rule for knowing, when we look at the smaller Diagram, what Attributes belong to the things in each compartment.

8. What does "some" mean in Logic?
[See pp. 55, 6]

9. In what sense do we use the word 'Universe' in this Game?

10. What is a 'Double' Proposition? Give examples.

11. When is a class of Things said to be 'exhaustively' divided? Give examples.

12. Explain the phrase "sitting on the fence."

13. What two partial Propositions make up, when taken together, "all x are y"?

14. What are 'Individual' Propositions? Give examples.

15. What kinds of Propositions imply, in this Game, the *existence* of their Subjects?

16. When a Proposition contains more than two Attributes, these Attributes may in some cases be re-arranged, and shifted from one Term to the other. In what cases may this be done? Give examples.

Break up each of the following into two *partial* Propositions:

17. All tigers are fierce.

18. All hard-boiled eggs are unwholesome.

19. I am happy.

20. John is not at home.

[See pp. 56, 7]

21. Give a rule for knowing, when we look at the larger Diagram, what Attributes belong to the Things contained in each compartment.

22. Explain 'Premisses', 'Conclusion', and 'Syllogism'. Give examples.

23. Explain the phrases 'Middle Term' and 'Middle Terms'.

24. In marking a pair of Premisses on the larger Diagram, why is it best to mark *negative* Propositions before *affirmative* ones?

25. Why is it of no consequence to us, as Logicians, whether the Premisses are true or false?

26. How can we work Syllogisms in which we are told that "some x are y" is to be understood to mean "the Attributes x, y are *compatible*", and "no x are y" to mean "the Attributes x, y are *incompatible*"?

27. What are the two kinds of 'Fallacies'?

28. How may we detect 'Fallacious Premisses'?

29. How may we detect a 'Fallacious Conclusion'?

30. Sometimes the Conclusion, offered to us, is not identical with the correct Conclusion, and yet cannot be fairly called 'Fallacious'. When does this happen? And what name may we give to such a Conclusion?

[See pp. 57—59]

§ 2. *Half of Smaller Diagram.*
Propositions to be represented.

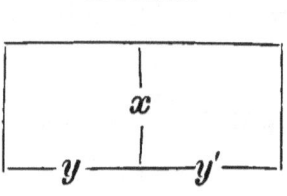

1. Some x are not-y.
2. All x are not-y.
3. Some x are y, and some are not-y.
4. No x exist.
5. Some x exist.
6. No x are not-y.
7. Some x are not-y, and some x exist.

Taking x = "judges"; y = "just";

8. No judges are just.
9. Some judges are unjust.
10. All judges are just.

Taking x = "plums"; y = "wholesome";

11. Some plums are wholesome.
12. There are no wholesome plums.
13. Plums are some of them wholesome, and some not.
14. All plums are unwholesome.

[See pp. 59, 60]

§ 2.]   CROSS QUESTIONS.   41

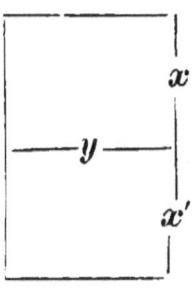

Taking $y=$ "diligent students"; $x=$ "successful";

15. No diligent students are unsuccessful.
16. All diligent students are successful.
17. No students are diligent.
18. There are some diligent, but unsuccessful, students.
19. Some students are diligent.

[See pp. 60, 1]

§ 3. *Half of Smaller Diagram.*
Symbols to be interpreted.

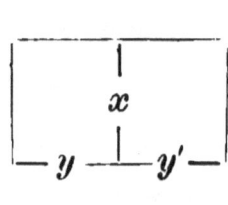

1. | 0 | 2. | 0 | 0 |

3. | — | 4. | 0 | 1 |

Taking $x=$ "good riddles"; $y=$ "hard";

5. | 1 | | 6. | 1 | 0 |

7. | 0 | 0 | 8. | 0 | |

[See pp. 61, 2]

[§ 3.]   *CROSS QUESTIONS.*   43

Taking $x=$ "lobsters"; $y=$ "selfish";

9. | | 1 | 10. | 0 | |

11. | 0 | 1 | 12. | 1 | 1 |

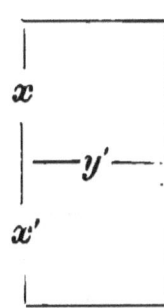

Taking $y=$ "healthy people"; $x=$ "happy";

13. | 0 |
    | 1 |

14. | −1 |

15. | 1 |
    | 1 |

16. | 0 |
    | 0 |

[See p. 62]

§ 4. *Smaller Diagram.*
Propositions to be represented.

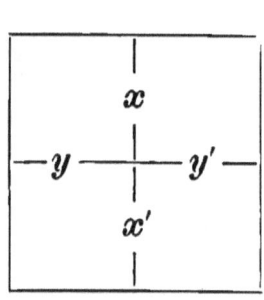

1. All *y* are *x*.
2. Some *y* are not-*x*.
3. No not-*x* are not-*y*.
4. Some *x* are not-*y*.
5. Some not-*y* are *x*.
6. No not-*x* are *y*.
7. Some not-*x* are not-*y*.
8. All not-*x* are not-*y*.
9. Some not-*y* exist.
10. No not-*x* exist.
11. Some *y* are *x*, and some are not-*x*.
12. All *x* are *y*, and all not-*y* are not-*x*.

[See pp. 62, 3]

Taking "nations" as Universe; $x=$ "civilised";
$y=$ "warlike";

13. No uncivilised nation is warlike.
14. All unwarlike nations are uncivilised.
15. Some nations are unwarlike.
16. All warlike nations are civilised, and all civilised nations are warlike.
17. No nation is uncivilised.

Taking "crocodiles" as Universe; $x=$ "hungry"; and $y=$ "amiable";

18. All hungry crocodiles are unamiable.
19. No crocodiles are amiable when hungry.
20. Some crocodiles, when not hungry, are amiable; but some are not.
21. No crocodiles are amiable, and some are hungry.
22. All crocodiles, when not hungry, are amiable; and all unamiable crocodiles are hungry.
23. Some hungry crocodiles are amiable, and some that are not hungry are unamiable.

[See pp. 63, 4]

§ 5. *Smaller Diagram.*
Symbols to be interpreted.

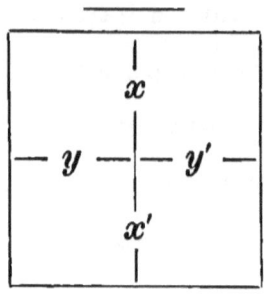

1.
2.
3.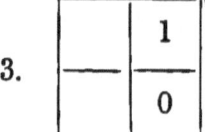
4.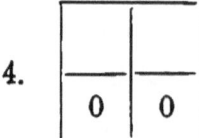

Taking "houses" as Universe; $x=$ "built of brick"; and $y=$ "two-storied"; interpret

5.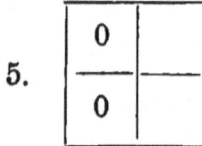
6.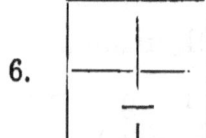
7.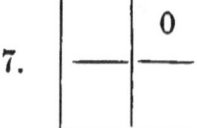
8.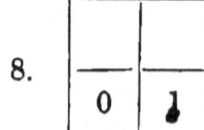

[See p. 65

[§ 5.] CROSS QUESTIONS. 47

Taking "boys" as Universe; $x=$ "fat";
and $y=$ "active"; interpret

9.
```
| 1 | 1 |
|---|---|
|   |   |
```

10.
```
|   | 0 |
|---|---|
|   | 1 |
```

11.
```
| 0 | 1 |
|---|---|
|   | 0 |
```

12.
```
| 1 |   |
|---|---|
| 0 | 1 |
```

Taking "cats" as Universe; $x=$ "green-eyed";
and $y=$ "good-tempered"; interpret

13.
```
| 0 | 0 |
|---|---|
|   | 0 |
```

14.
```
|   | 1 |
|---|---|
| 1 |   |
```

15.
```
| 1 |   |
|---|---|
|   | 0 |
```

16.
```
| 0 | 1 |
|---|---|
| 1 | 0 |
```

[See pp. 65, 6]

§ 6. *Larger Diagram.*
Propositions to be represented.

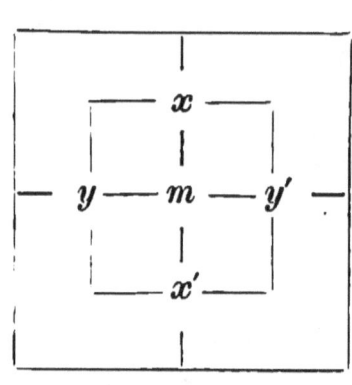

1. No x are m.
2. Some y are m'.
3. All m are x'.
4. No m' are y'.
5. No m are x; } All y are m.
6. Some x are m; } No y are m.
7. All m are x'; } No m are y.
8. No x' are m; } No y' are m'.

[See pp. 67, 8]

[§ 6.] CROSS QUESTIONS. 49

Taking "rabbits" as Universe; $m=$ "greedy";
$x=$ "old"; and $y=$ "black"; represent

9. No old rabbits are greedy.
10. Some not-greedy rabbits are black.
11. All white rabbits are free from greediness.
12. All greedy rabbits are young.
13. No old rabbits are greedy; $\Big\}$
    All black rabbits are greedy.
14. All rabbits, that are not greedy, are black; $\Big\}$
    No old rabbits are free from greediness.

Taking "birds" as Universe; $m=$ "that sing loud";
$x=$ "well-fed"; and $y=$ "happy"; represent

15. All well-fed birds sing loud; $\Big\}$
    No birds, that sing loud, are unhappy.
16. All birds, that do not sing loud, are unhappy; $\Big\}$
    No well-fed birds fail to sing loud.

Taking "persons" as Universe; $m=$ "in the house";
$x=$ "John"; and $y=$ "having a tooth-ache"; represent

17. John is in the house; $\Big\}$
    Everybody in the house is suffering from tooth-ache.
18. There is no one in the house but John; $\Big\}$
    Nobody, out of the house, has a tooth-ache.

[See pp. 68—70]

E

Taking "persons" as Universe; $m=$ "I";
$x=$ "that has taken a walk"; $y=$ "that feels better";
represent

19. I have been out for a walk;
 I feel much better.

Choosing your own 'Universe' &c., represent

20. I sent him to bring me a kitten;
 He brought me a kettle by mistake.

[See pp. 70, 1]

§ 7. *Both Diagrams to be employed.*

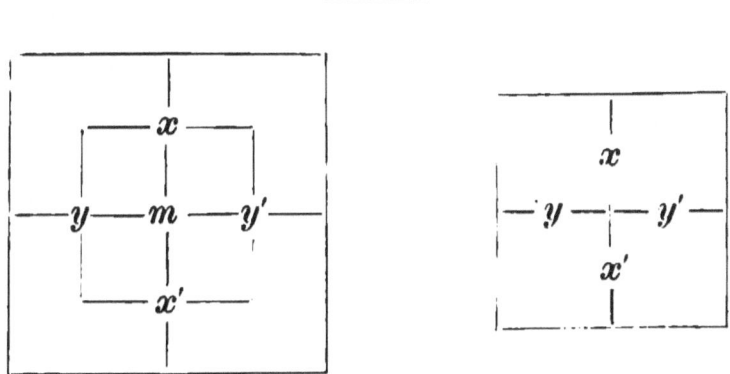

N.B. In each Question, a small Diagram should be drawn, for x and y only, and marked in accordance with the given large Diagram: and then as many Propositions as possible, for x and y, should be read off from this small Diagram.

1.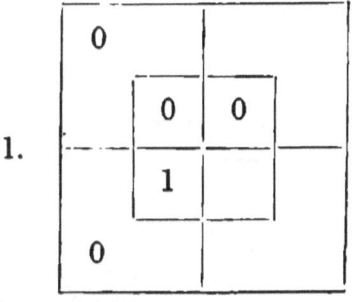

2.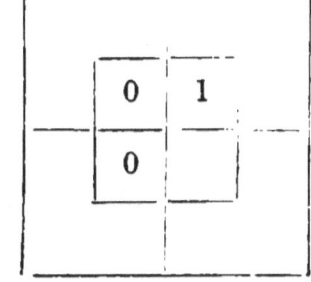

[See p. 72]

3. 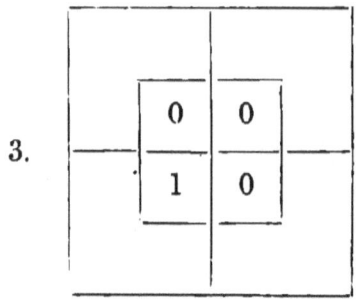 4.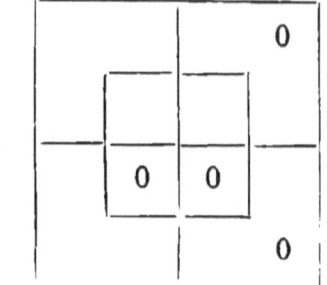

Mark, on a large Diagram, the following pairs of Propositions from the preceding Section: then mark a small Diagram in accordance with it, &c.

    5. No. 13. [see p. 49]   9. No. 17.
    6. No. 14.   10. No. 18.
    7. No. 15.   11. No. 19. [see p. 50]
    8. No. 16.   12. No. 20.

Mark, on a large Diagram, the following Pairs of Propositions: then mark a small Diagram, &c. These are, in fact, Pairs of *Premisses* for Syllogisms: and the results, read off from the small Diagram, are the *Conclusions*.

    13. No exciting books suit feverish patients;
        Unexciting books make one drowsy.

    14. Some, who deserve the fair, get their deserts;
        None but the brave deserve the fair.

    15. No children are patient;
        No impatient person can sit still.

[See pp. 72—5]

16. All pigs are fat ;
    No skeletons are fat.

17. No monkeys are soldiers ;
    All monkeys are mischievous.

18. None of my cousins are just ;
    No judges are unjust.

19. Some days are rainy ;
    Rainy days are tiresome.

20. All medicine is nasty ;
    Senna is a medicine.

21. Some Jews are rich ;
    All Patagonians are Gentiles.

22. All teetotalers like sugar ;
    No nightingale drinks wine.

23. No muffins are wholesome ;
    All buns are unwholesome.

24. No fat creatures run well ;
    Some greyhounds run well.

25. All soldiers march ;
    Some youths are not soldiers.

26. Sugar is sweet ;
    Salt is not sweet.

27. Some eggs are hard-boiled ;
    No eggs are uncrackable.

28. There are no Jews in the house ;
    There are no Gentiles in the garden.

[See pp. 75—82]

29. All battles are noisy ;
    What makes no noise may escape notice.

30. No Jews are mad ;
    All Rabbis are Jews.

31. There are no fish that cannot swim ;
    Some skates are fish.

32. All passionate people are unreasonable ;
    Some orators are passionate.

[See pp. 82—84]

CHAPTER III.

CROOKED ANSWERS.

" I answered him, as I thought good,
'As many as red-herrings grow in the wood '."

---

§ 1. *Elementary.*

1. Whatever can be "attributed to", that is "said to belong to", a Thing, is called an 'Attribute'. For example, "baked", which can (frequently) be attributed to "Buns", and "beautiful", which can (seldom) be attributed to "Babies".

2. When they are the Names of two Things (for example, "these Pigs are fat Animals"), or of two Attributes (for example, "pink is light red").

3. When one is the Name of a Thing, and the other the Name of an Attribute (for example, "these Pigs are pink"), since a Thing cannot actually *be* an Attribute.

4. That the Substantive shall be supposed to be repeated at the end of the sentence (for example, "these Pigs are pink (Pigs)").

5. A 'Proposition' is a sentence stating that some, or none, or all, of the Things belonging to a certain class,
[See p. 37]

called the 'Subject', are also Things belonging to a certain other class, called the 'Predicate'. For example, "some new Cakes are not nice", that is (written in full) "some new Cakes are not nice Cakes"; where the class "new Cakes" is the Subject, and the class "not-nice Cakes" is the Predicate.

6. A Proposition, stating that *some* of the Things belonging to its Subject are so-and-so, is called 'Particular'. For example, "some new Cakes are nice", "some new Cakes are not nice."

A Proposition, stating that *none* of the Things belonging to its Subject, or that *all* of them, are so-and-so, is called 'Universal'. For example, "no new Cakes are nice", "all new Cakes are not nice".

7. The Things in each compartment possess *two* Attributes, whose symbols will be found written on two of the *edges* of that compartment.

8. "One or more."

9. As a name of the class of Things to which the whole Diagram is assigned.

10. A Proposition containing two statements. For example, "some new Cakes are nice and some are not-nice."

11. When the whole class, thus divided, is "exhausted" among the sets into which it is divided, there being no member of it which does not belong to some one of them. For example, the class "new Cakes" is "exhaustively"

[See pp. 37, 8]

§ 1.] *ELEMENTARY.* 57

divided into "nice" and "not-nice" since *every* new Cake must be one or the other.

12. When a man cannot make up his mind which of two parties he will join, he is said to be "sitting on the fence"——not being able to decide on which side he will jump down.

13. "Some x are y" and "no x are y'".

14. A Proposition, whose Subject is a single Thing, is called 'Individual'. For example, "I am happy", "John is not at home". These are Universal Propositions, being the same as "all the I's that exist are happy", "*all* the Johns, that I am now considering, are not at home".

15. Propositions beginning with "some" or "all".

16. When they begin with "some" or "no". For example, "some abc are def" may be re-arranged as "some bf are $acde$", each being equivalent to "some $abcdef$ exist".

17. Some tigers are fierce,
    No tigers are not-fierce.

18. Some hard-boiled eggs are unwholesome,
    No hard-boiled eggs are wholesome.

19. Some I's are happy,
    No I's are unhappy.

20. Some Johns are not at home,
    No Johns are at home.

21. The Things, in each compartment of the larger Diagram, possess *three* Attributes, whose symbols will be [See pp. 38, 9.]

found written at three of the *corners* of the compartment (except in the case of m', which is not actually inserted in the Diagram, but is *supposed* to stand at each of its four outer corners).

22. If the Universe of Things be divided with regard to three different Attributes; and if two Propositions be given, containing two different couples of these Attributes; and if from these we can prove a third Proposition, containing the two Attributes that have not yet occurred together; the given Propositions are called 'the Premisses', the third one 'the Conclusion', and the whole set 'a Syllogism'. For example, the Premisses might be "no m are x'" and "all m' are y"; and it might be possible to prove from them a Conclusion containing x and y.

23. If an Attribute occurs in both Premisses, the Term containing it is called 'the Middle Term'. For example, if the Premisses are "some m are x" and "no m are y'", the class of "m-Things" is 'the Middle Term.'

If an Attribute occurs in one Premiss, and its contradictory in the other, the Terms containing them may be called 'the Middle Terms'. For example, if the Premisses are "no m are x'" and "all m' are y", the two classes of "m-Things" and "m'-Things" may be called 'the Middle Terms'.

24. Because they can be marked with *certainty*: whereas *affirmative* Propositions (that is, those that begin with "some" or "all") sometimes require us to place a red counter 'sitting on a fence'.

[See p. 39]

§ 1.]   *ELEMENTARY.*   59

25. Because the only question we are concerned with is whether the Conclusion *follows logically* from the Premisses, so that, if *they* were true, *it* also would be true.

26. By understanding a red counter to mean "this compartment *can* be occupied", and a grey one to mean "this compartment *cannot* be occupied" or "this compartment *must* be empty".

27. 'Fallacious Premisses' and 'Fallacious Conclusion'.

28. By finding, when we try to transfer marks from the larger Diagram to the smaller, that there is 'no information' for any of its four compartments.

29. By finding the correct Conclusion, and then observing that the Conclusion, offered to us, is neither identical with it nor a part of it.

30. When the offered Conclusion is *part* of the correct Conclusion. In this case, we may call it a 'Defective Conclusion'.

§ 2. *Half of Smaller Diagram.*
Propositions represented.

1. | | 1 | 2. | 0 | 1 |

3. | 1 | 1 | 4. | 0 | 0 |

[See pp. 39, 40]

5. | — | 6. | 0 |

7. | | 1 | It might be thought that the proper Diagram would be | — 1 |, in order to express " some x exist " : but this is really contained in " some x are y'." To put a red counter on the division-line would only tell us " *one of the two* compartments is occupied ", which we know already, in knowing that *one* is occupied.

8. No x are y.   i.e. | 0 | |

9. Some x are y'.   i.e. | | 1 |

10. All x are y.   i.e. | 1 | 0 |

11. Some x are y.   i.e. | 1 | |

12. No x are y.   i.e. | 0 | |

13. Some x are y, and some are y'.   i.e. | 1 | 1 |

14. All x are y'.   i.e. | 0 | 1 |

15. No y are x'.   i.e. | | / 0 |

[See pp. 40, 1]

§ 2.]   PROPOSITIONS REPRESENTED.   61

16. All *y* are *x*.   i. e.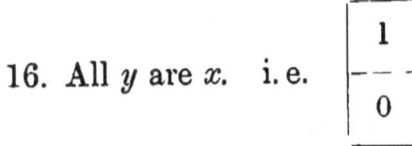

17. No *y* exist.   i. e.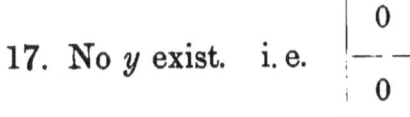

18. Some *y* are *x'*.   i. e.

19. Some *y* exist.   i. e.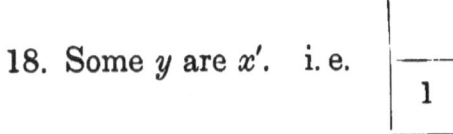

§ 3. *Half of Smaller Diagram.*
Symbols interpreted.

1. No *x* are *y'*.
2. No *x* exist.
3. Some *x* exist.
4. All *x* are *y'*.
5. Some *x* are *y*.   i. e. Some good riddles are hard.
6. All *x* are *y*.   i. e. All good riddles are hard.
7. No *x* exist.   i. e. No riddles are good.

[See pp. 41, 2]

8. No x are y. i. e. No good riddles are hard.
9. Some x are y'. i. e. Some lobsters are unselfish.
10. No x are y. i. e. No lobsters are selfish.
11. All x are y'. i. e. All lobsters are unselfish.
12. Some x are y, and some are y'. i. e. Some lobsters are selfish, and some are unselfish.
13. All y' are x'. i. e. All invalids are unhappy.
14. Some y' exist. i. e. Some people are unhealthy.
15. Some y' are x, and some are x'. i. e. Some invalids are happy, and some are unhappy.
16. No y' exist. i. e. Nobody is unhealthy.

§ 4. *Smaller Diagram.*
Propositions represented.

1. [diagram: 1 in upper-left, 0 in lower-left]
2. [diagram: 1 in lower-left]
3. [diagram: 0 in lower-right]
4. [diagram: 1 in upper-right, • in lower-left]

[See pp. 42—4]

§ 4.] PROPOSITIONS REPRESENTED. 63

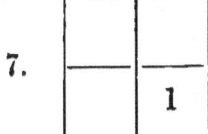

5. 6.

7. 8.

9. 10.

11. 12.

13. No x' are y. i.e.

14. All y' are x'. i.e.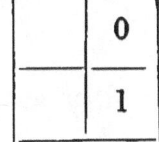

15. Some y' exist. i.e.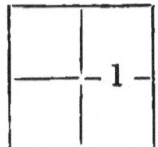

[See pp. 44, 5]

64   *CROOKED ANSWERS.*   [Ch. III. § 4.

16. All y are x, and all x are y.   i. e.

1	0
0	

17. No x' exist.   i. e.

0	0

18. All x are y'.   i. e.

0	1

19. No x are y.   i. e.

0	

20. Some x' are y, and some are y'.   i. e.

1	1

21. No y exist, and some x exist.   i. e.

0	1
0	

22. All x' are y, and all y' are x.   i. e.

	1
1	0

23. Some x are y, and some x' are y'.   i. e.

1	
	1

[See p. 45]

§ 5. *Smaller Diagram.*
Symbols interpreted.

1. Some y are not-x,
   or, Some not-x are y.
2. No not-x are not-y,
   or, No not-y are not-x.
3. No not-y are x.
4. No not-x exist. i.e. No Things are not-x.
5. No y exist. i.e. No houses are two-storied.
6. Some x' exist. i.e. Some houses are not built of brick.
7. No x are y'. Or, no y' are x. i.e. No houses, built of brick, are other than two-storied. Or, no houses, that are not two-storied, are built of brick.
8. All x' are y'. i.e. All houses, that are not built of brick, are not two-storied.
9. Some x are y, and some are y'. i.e. Some fat boys are active, and some are not.
10. All y' are x'. i.e. All lazy boys are thin.
11. All x are y', and all y' are x. i.e. All fat boys are lazy, and all lazy ones are fat.

[See pp. 46, 7]

12. All y are x, and all x' are y.   i.e. All active boys are fat, and all thin ones are lazy.

13. No x exist, and no y' exist.   i.e. No cats have green eyes, and none have bad tempers.

14. Some x are y', and some x' are y. Or, some y are x', and some y' are x.   i.e. Some green-eyed cats are bad-tempered, and some, that have not green eyes, are good-tempered. Or, some good-tempered cats have not green eyes, and some bad-tempered ones have green eyes.

15. Some x are y, and no x' are y'. Or, some y are x, and no y' are x'. i.e. Some green-eyed cats are good-tempered, and none, that are not green-eyed, are bad-tempered. Or, some good-tempered cats have green eyes, and none, that are bad-tempered, have not green eyes.

16. All x are y', and all x' are y. Or, all y are x', and all y' are x. i.e. All green-eyed cats are bad-tempered, and all, that have not green eyes, are good-tempered. Or, all good-tempered ones have eyes that are not green, and all bad-tempered ones have green eyes.

[See p. 47]

[Ch. III. § 6.]   CROOKED ANSWERS.   67

§ 6. *Larger Diagram.*
Propositions represented.

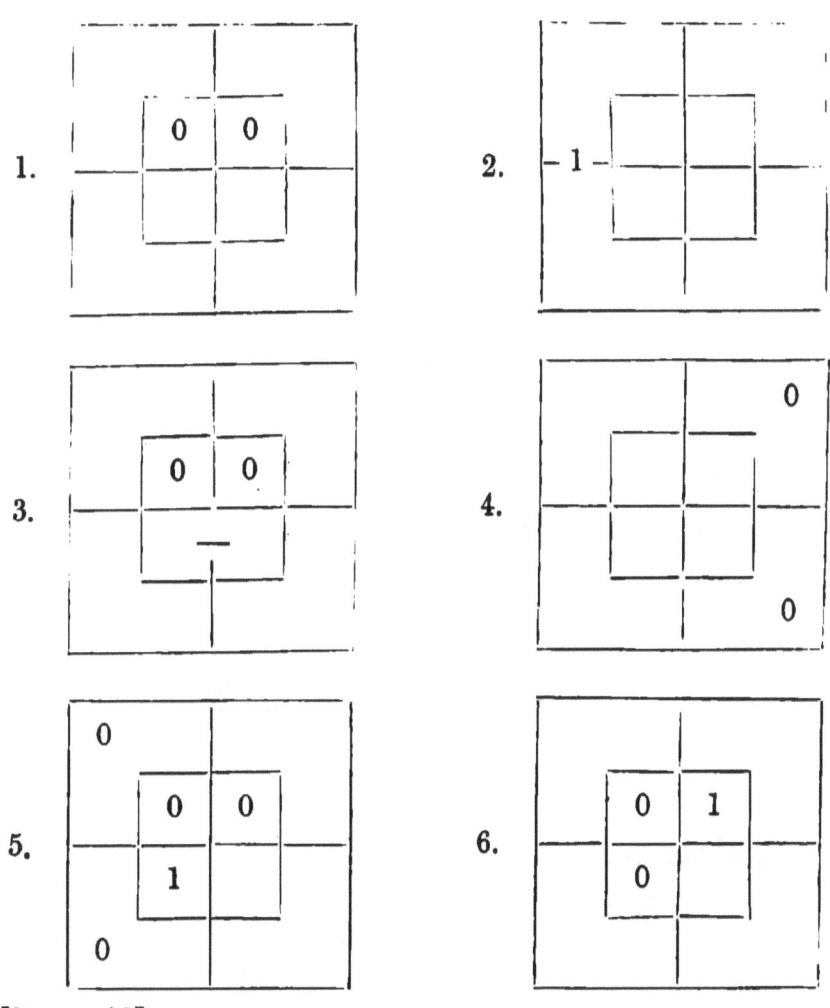

[See p. 48]

F 2

7. 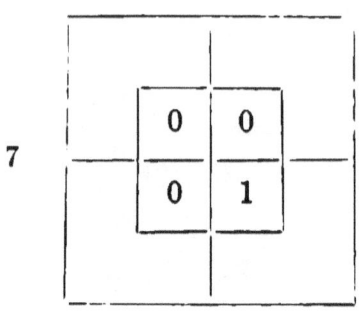   8.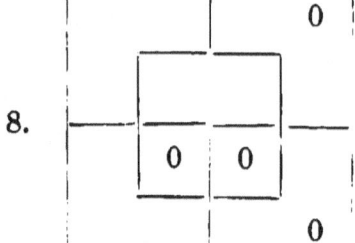

9. No x are m.   i.e.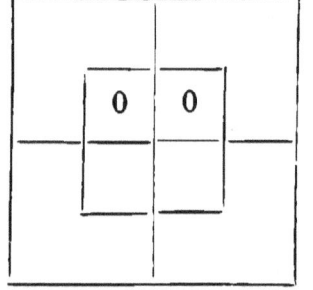

10. Some m' are y.   i.e.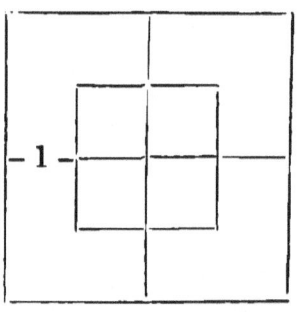

11. All y' are m'.   i.e.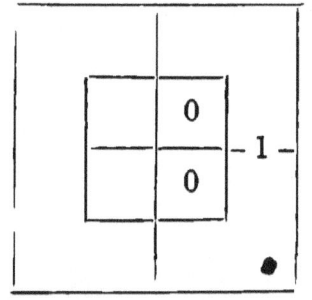

[See pp. 48, 9]

12. All m are x'.   i. e.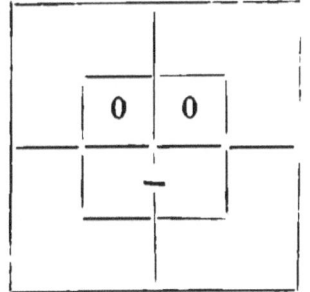

13. No x are m;
    All y are m. } i. e.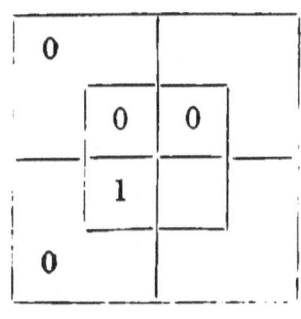

14. All m' are y;
    No x are m'. } i. e.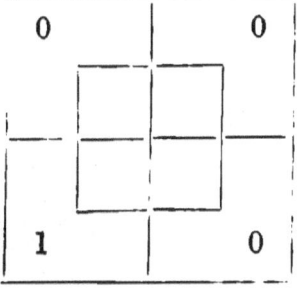

15. All x are m;
    No m are y'. } i. e.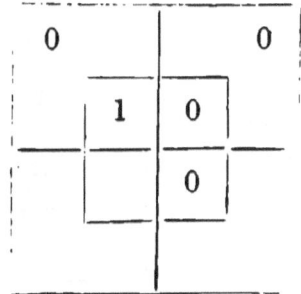

[See p. 49]

16. All m' are y' ;
No x are m'. } i. e.

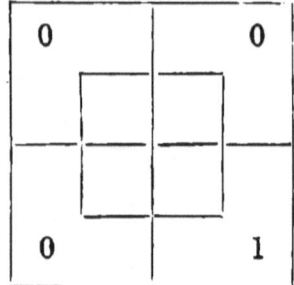

17. All x are m ;
All m are y. } i. e.

[See remarks on No. 7, p. 60.]

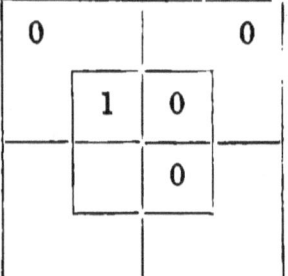

18. No x' are m ;
No m' are y. } i. e.

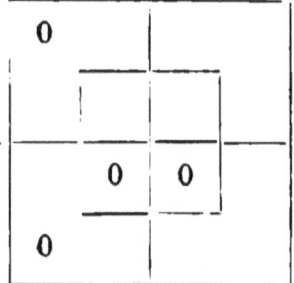

19. All m are x ;
All m are y. } i. e.

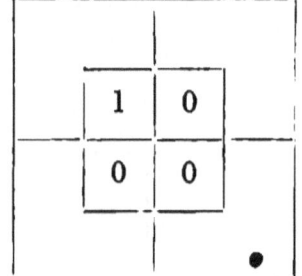

[See pp. 49, 50]

§ 6.]   PROPOSITIONS REPRESENTED.   71

20. We had better take "persons" as Universe. We may choose "myself" as 'Middle Term', in which case the Premisses will take the form

I am a-person-who-sent-him-to-bring-a-kitten ; }
I am a-person-to-whom-he-brought-a-kettle-by-mistake. }

Or we may choose "he" as 'Middle Term', in which case the Premisses will take the form

He is a-person-whom-I-sent-to-bring-me-a-kitten ; }
He is a-person-who-brought-me-a-kettle-by-mistake. }

The latter form seems best, as the interest of the anecdote clearly depends on *his* stupidity——not on what happened to *me*. Let us then make m = "he"; x = "persons whom I sent, &c."; and y = "persons who brought, &c."

Hence, All m are x; } and the required Diagram is
       All m are y. }

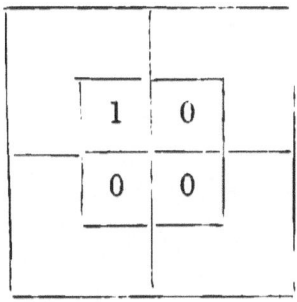

[See p. 50]

§ 7. *Both Diagrams employed.*

1. 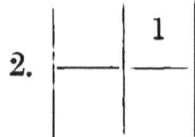 i.e. All y are x'.

2. 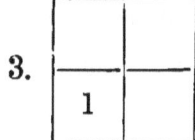 i.e. Some x are y'; or, Some y' are x.

3. 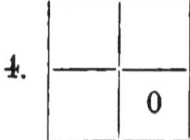 i.e. Some y are x'; or, Some x' are y.

4. 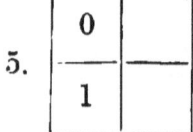 i.e. No x' are y'; or, No y' are x'.

5. 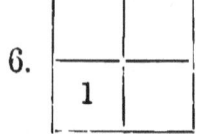 i.e. All y are x'.   i.e. All black rabbits are young.

6. 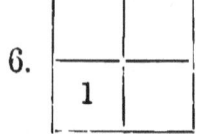 i.e. Some y are x'.   i.e. Some black rabbits are young.

[See pp. 51, 2]

§ 7.] BOTH DIAGRAMS EMPLOYED.

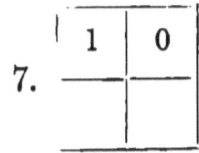
7. i.e. All x are y. i.e. All well-fed birds are happy.

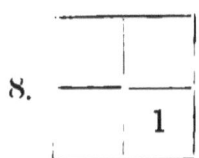
8. i.e. Some x' are y'. i.e. Some birds, that are not well-fed, are unhappy; or, Some unhappy birds are not well-fed.

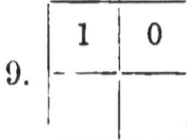
9. i.e. All x are y. i.e. John has got a tooth-ache.

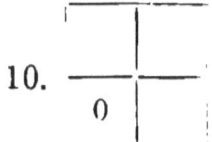
10. i.e. No x' are y. i.e. No one, but John, has got a tooth-ache.

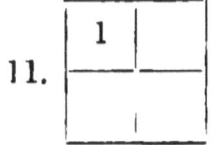
11. i.e. Some x are y. i.e. Some one, who has taken a walk, feels better.

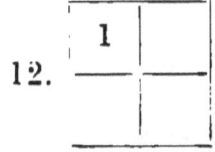
12. i.e. Some x are y. i.e. Some one, whom I sent to bring me a kitten, brought me a kettle by mistake.

[See p. 52]

13.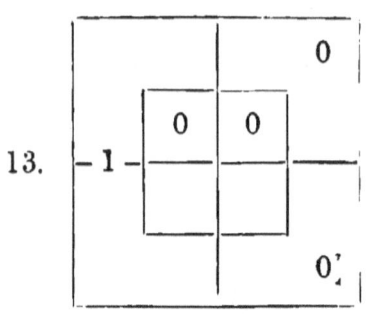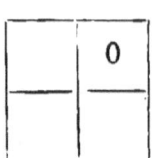

Let "books" be Universe; m = "exciting",
x = "that suit feverish patients"; y = "that make one drowsy".

No m are x;
All m' are y. } ∴ No y' are x.

i. e. No books suit feverish patients, except such as make one drowsy.

14.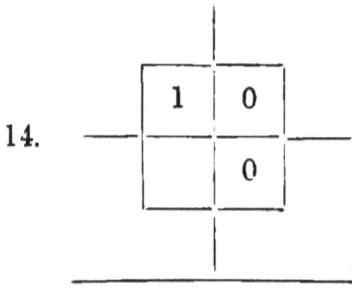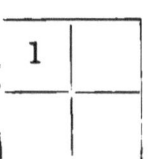

Let "persons" be Universe; m = "that deserve the fair":
x = "that get their deserts"; y = "brave".

Some m are x;
No y' are m. } ∴ Some y are x.

i. e. Some brave persons get their deserts.

[See p. 52]

15.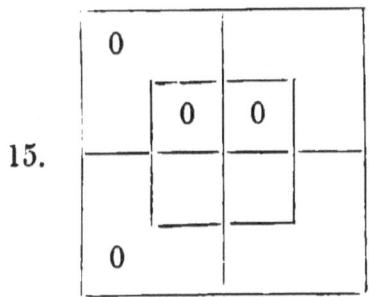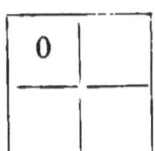

Let "persons" be Universe ; $m =$ "patient" ;
$x =$ "children" ; $y =$ "that can sit still".

No x are m ; }
No m' are y. } ∴ No x are y.

    i. e. No children can sit still.

16.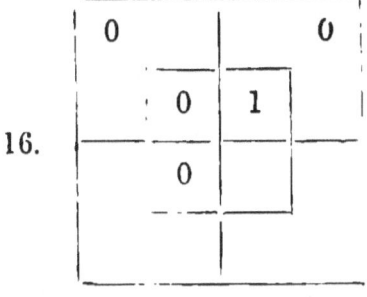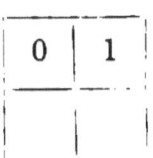

Let "things" be Universe ; $m =$ "fat" ; $x =$ "pigs" ;
$y =$ "skeletons".

All x are m ; }
No y are m. } ∴ All x are y'.

    i. e. All pigs are not-skeletons.

[See pp. 52, 3]

17.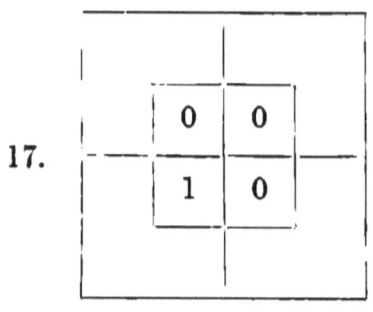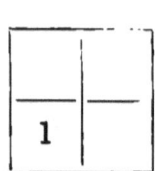

Let " creatures " be Universe; m = " monkeys ";
x = " soldiers "; y = " mischievous ".

No m are x; }
All m are y. } ∴ Some y are x'.

i. e. Some mischievous creatures are not soldiers.

18.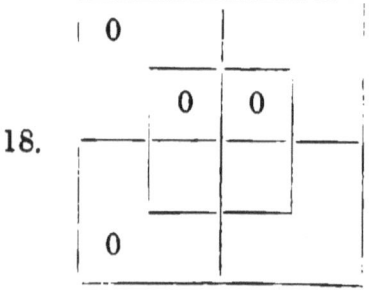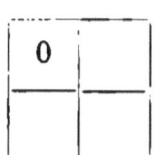

Let " persons " be Universe; m = " just ";
x = " my cousins "; y = " judges ".

No x are m; }
No y are m'. } ∴ No x are y.

i. e. None of my cousins are judges.

[See p. 53]

BOTH DIAGRAMS EMPLOYED.

19.

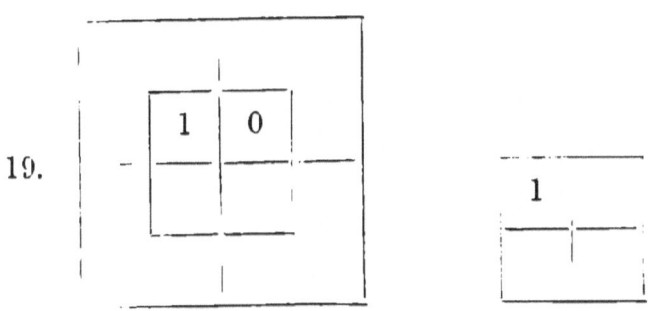

Let "periods" be Universe; $m =$ "days";
$x =$ "rainy"; $y =$ "tiresome".

Some m are x; }
All xm are y. } ∴ Some x are y.

i. e. Some rainy periods are tiresome.

N.B. These are not legitimate Premisses, since the Conclusion is really part of the second Premiss, so that the first Premiss is superfluous. This may be shown, in letters, thus:—

"All xm are y" contains "Some xm are y", which contains "Some x are y". Or, in words, "All rainy days are tiresome" contains "Some rainy days are tiresome", which contains "Some rainy periods are tiresome".

Moreover, the first Premiss, besides being superfluous, is actually contained in the second; since it is equivalent to "Some rainy days exist", which, as we know, is implied in the Proposition "All rainy days are tiresome".

Altogether, a *most* unsatisfactory Pair of Premisses! [See p. 53]

20.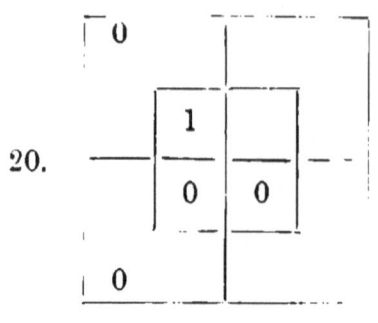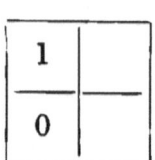

Let "things" be Universe; m = "medicine";
x = "nasty"; y = "senna".

All m are x;
All y are m. } ∴ All y are x.

i. e. Senna is nasty.

[See remarks on No. 7, p. 60.]

21.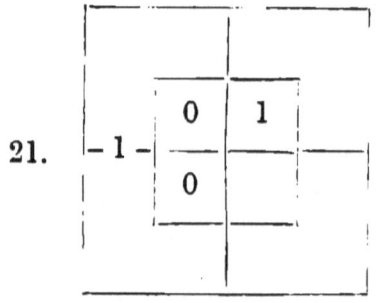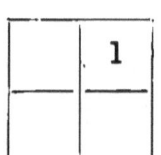

Let "persons" be Universe; m = "Jews";
x = "rich"; y = "Patagonians".

Some m are x;
All y are m'. } ∴ Some x are y'.

i. e. Some rich persons are not Patagonians.

[See p. 53]

22.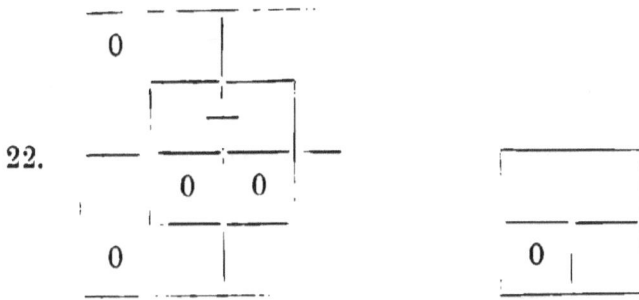

Let "creatures" be Universe; m = "teetotalers";
x = "that like sugar"; y = "nightingales".
All m are x;
No y are m'. $\}$ ∴ No y are x'.
i.e. No nightingales dislike sugar.

23.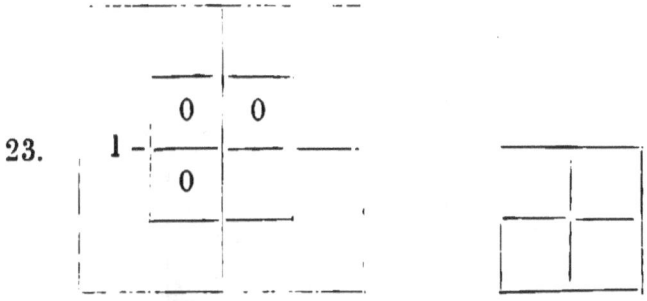

Let "food" be Universe; m = "wholesome";
x = "muffins"; y = "buns".
No x are m;
All y are m. $\}$

There is 'no information' for the smaller Diagram; so no Conclusion can be drawn.
[See p. 53]

80   *CROOKED ANSWERS.*   [Ch. III.

24.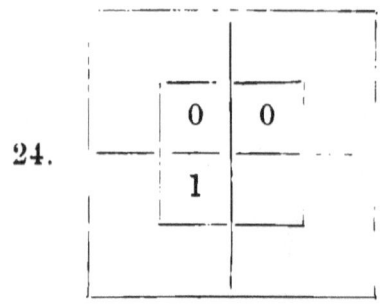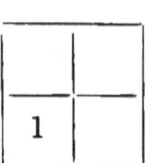

Let " creatures" be Universe; m = " that run well ";
x = " fat"; y = " greyhounds ".

No x are m;
Some y are m. } ∴ Some y are x'.

i. e. Some greyhounds are not fat.

25.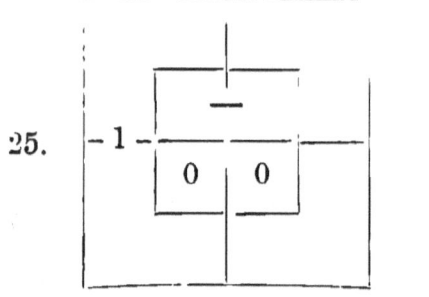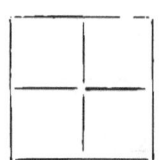

Let " persons " be Universe; m = " soldiers ";
x = " that march "; y = " youths ".

All m are x;
Some y are m'. }

There is 'no information' for the smaller Diagram; so no Conclusion can be drawn.

[See p. 53]

26.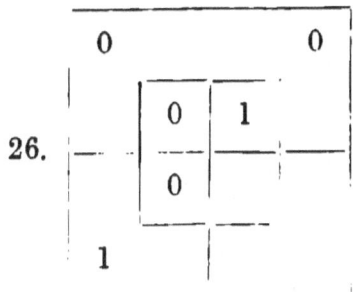

Let "food" be Universe; m = "sweet";
x = "sugar"; y = "salt".
All x are m; }   { All x are y'.
All y are m'. } ∴ { All y are x'.

i. e. { Sugar is not salt.
{ Salt is not sugar.

27.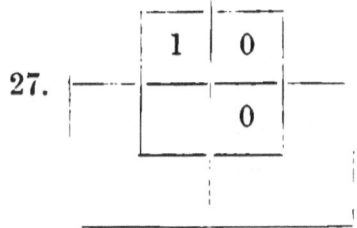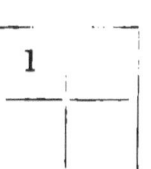

Let "Things" be Universe; m = "eggs";
x = "hard-boiled"; y = "crackable".
Some m are x; }
No m are y'. } ∴ Some x are y.

i. e. Some hard-boiled things can be cracked.
[See p. 53]

G

28.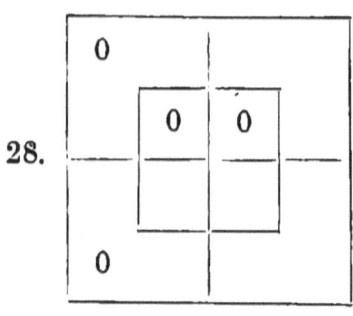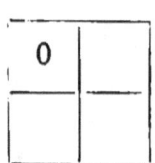

Let "persons" be Universe; m = "Jews"; x = "that are in the house"; y = "that are in the garden".

No m are x;
No m' are y. $\Big\}$ ∴ No x are y.

i. e. No persons, that are in the house, are also in the garden.

29.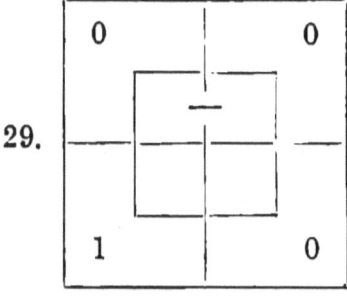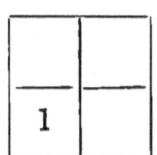

Let "Things" be Universe; m = "noisy"; x = "battles"; y = "that may escape notice".

All x are m;
All m' are y. $\Big\}$ ∴ Some x' are y.

i. e. Some things, that are not battles, may escape notice.

[See pp. 53, 54]

§ 7.] BOTH DIAGRAMS EMPLOYED.

30.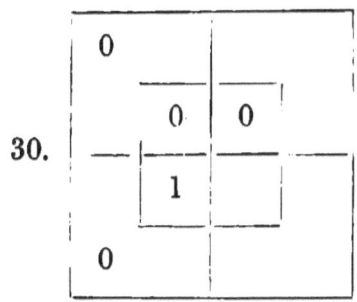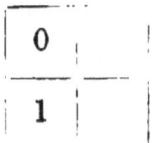

Let "persons" be Universe; m = "Jews"; x = "mad"; y = "Rabbis".

No m are x; } ∴ All y are x'.
All y are m.

i.e. All Rabbis are sane.

31.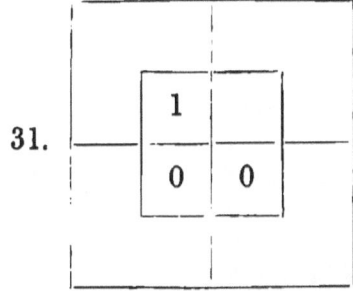

Let "Things" be Universe; m = "fish"; x = "that can swim"; y = "skates".

No m are x'; } ∴ Some y are x.
Some y are m.

i.e. Some skates can swim.

[See p. 54]

32.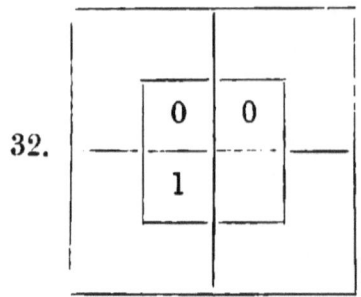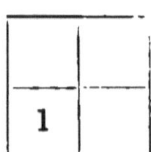

Let "people" be Universe; m = "passionate"; x = "reasonable"; y = "orators".

All m are x';
Some y are m. } ∴ Some y are x'.

i. e. Some orators are unreasonable.

[See remarks on No. 7, p. 60.]

[See p. 54]

CHAPTER IV.

HIT OR MISS.

"Thou canst not hit it, hit it, hit it,
Thou canst not hit it, my good man."

---

1. Pain is wearisome;
   No pain is eagerly wished for.
2. No bald person needs a hair-brush;
   No lizards have hair.
3. All thoughtless people do mischief;
   No thoughtful person forgets a promise.
4. I do not like John;
   Some of my friends like John.
5. No potatoes are pine-apples;
   All pine-apples are nice.
6. No pins are ambitious;
   No needles are pins.
7. All my friends have colds;
   No one can sing who has a cold.
8. All these dishes are well-cooked;
   Some dishes are unwholesome if not well-cooked.

9. No medicine is nice ;
    Senna is a medicine.

10. Some oysters are silent ;
    No silent creatures are amusing.

11. All wise men walk on their feet ;
    All unwise men walk on their hands.

12. "Mind your own business ;
    This quarrel is no business of yours."

13. No bridges are made of sugar ;
    Some bridges are picturesque.

14. No riddles interest me that can be solved ;
    All these riddles are insoluble.

15. John is industrious ;
    All industrious people are happy.

16. No frogs write books ;
    Some people use ink in writing books.

17. No pokers are soft ;
    All pillows are soft.

18. No antelope is ungraceful ;
    Graceful animals delight the eye.

19. Some uncles are ungenerous ;
    All merchants are generous.

20. No unhappy people chuckle ;
    No happy people groan.

21. Audible music causes vibration in the air ;
    Inaudible music is not worth paying for.

22. He gave me five pounds;
    I was delighted.

23. No old Jews are fat millers;
    All my friends are old millers.

24. Flour is good for food;
    Oatmeal is a kind of flour.

25. Some dreams are terrible;
    No lambs are terrible.

26. No rich man begs in the street;
    All who are not rich should keep accounts.

27. No thieves are honest;
    Some dishonest people are found out.

28. All wasps are unfriendly;
    All puppies are friendly.

29. All improbable stories are doubted;
    None of these stories are probable.

30. "He told me you had gone away."
    "He never says one word of truth."

31. His songs never last an hour;
    A song, that lasts an hour, is tedious.

32. No bride-cakes are wholesome;
    Unwholesome food should be avoided.

33. No old misers are cheerful;
    Some old misers are thin.

34. All ducks waddle;
    Nothing that waddles is graceful.

35. No Professors are ignorant;
    Some ignorant people are conceited.

36. Toothache is never pleasant;
    Warmth is never unpleasant.

37. Bores are terrible;
    You are a bore.

38. Some mountains are insurmountable;
    All stiles can be surmounted.

39. No Frenchmen like plumpudding;
    All Englishmen like plumpudding.

40. No idlers win fame;
    Some painters are not idle.

41. No lobsters are unreasonable;
    No reasonable creatures expect impossibilities.

42. No kind deed is unlawful;
    What is lawful may be done without fear.

43. No fossils can be crossed in love;
    An oyster may be crossed in love.

44. "This is beyond endurance!"
    "Well, nothing beyond endurance has ever happened to *me*."

45. All uneducated men are shallow;
    All these students are educated.

46. All my cousins are unjust;
    No judges are unjust.

47. No country, that has been explored, is infested by dragons;
    Unexplored countries are fascinating.

48. No misers are generous;
    Some old men are not generous.

49. A prudent man shuns hyænas;
    No banker is imprudent.

50. Some poetry is original;
    No original work is producible at will.

51. No misers are unselfish;
    None but misers save egg-shells.

52. All pale people are phlegmatic;
    No one, who is not pale, looks poetical.

53. All spiders spin webs;
    Some creatures, that do not spin webs, are savage.

54. None of my cousins are just;
    All judges are just.

55. John is industrious;
    No industrious people are unhappy.

56. Umbrellas are useful on a journey;
    What is useless on a journey should be left behind.

57. Some pillows are soft;
    No pokers are soft.

58. I am old and lame;
    No old merchant is a lame gambler.

59. No eventful journey is ever forgotten;
    Uneventful journeys are not worth writing a book about.

60. Sugar is sweet;
    Some sweet things are liked by children.

61. Richard is out of temper;
    No one but Richard can ride that horse.

62. All jokes are meant to amuse;
    No Act of Parliament is a joke.

63. "I saw it in a newspaper."
    "All newspapers tell lies."

64. No nightmare is pleasant;
    Unpleasant experiences are not anxiously desired.

65. Prudent travellers carry plenty of small change;
    Imprudent travellers lose their luggage.

66. All wasps are unfriendly;
    No puppies are unfriendly.

67. He called here yesterday;
    He is no friend of mine.

68. No quadrupeds can whistle;
    Some cats are quadrupeds.

69. No cooked meat is sold by butchers;
    No uncooked meat is served at dinner.

70. Gold is heavy;
    Nothing but gold will silence him.

71. Some pigs are wild;
    There are no pigs that are not fat.

72. No emperors are dentists;
All dentists are dreaded by children.

73. All, who are not old, like walking;
Neither you nor I are old.

74. All blades are sharp;
Some grasses are blades.

75. No dictatorial person is popular;
She is dictatorial.

76. Some sweet things are unwholesome;
No muffins are sweet.

77. No military men write poetry;
No generals are civilians.

78. Bores are dreaded;
A bore is never begged to prolong his visit.

79. All owls are satisfactory;
Some excuses are unsatisfactory.

80. All my cousins are unjust;
All judges are just.

81. Some buns are rich;
All buns are nice.

82. No medicine is nice;
No pills are unmedicinal.

83. Some lessons are difficult;
What is difficult needs attention.

84. No unexpected pleasure annoys me;
Your visit is an unexpected pleasure.

85. Caterpillars are not eloquent;
    Jones is eloquent.

86. Some bald people wear wigs;
    All your children have hair.

87. All wasps are unfriendly;
    Unfriendly creatures are always unwelcome.

88. No bankrupts are rich;
    Some merchants are not bankrupts.

89. Weasels sometimes sleep;
    All animals sometimes sleep.

90. Ill-managed concerns are unprofitable;
    Railways are never ill-managed.

91. Everybody has seen a pig;
    Nobody admires a pig.

Extract a Pair of Premisses out of each of the following: and deduce the Conclusion, if there is one :—

92. "The Lion, as any one can tell you who has been chased by them as often as *I* have, is a very savage animal: and there are certain individuals among them, though I will not guarantee it as a general law, who do not drink coffee."

93. "It was most absurd of you to offer it! You might have known, if you had had any sense, that no old sailors ever like gruel!"

"But I thought, as he was an uncle of yours——"

"An uncle of mine, indeed! Stuff!"

"You may call it stuff, if you like. All I know is, *my* uncles are all old men: and they like gruel like anything!"

"Well, then *your* uncles are ——"

94. "Do come away! I can't stand this squeezing any more. No crowded shops are comfortable, you know very well."

"Well, who expects to be comfortable, out shopping?"

"Why, *I* do, of course! And I'm sure there are some shops, further down the street, that are not crowded. So ——"

95. "They say no doctors are metaphysical organists: and that lets me into a little fact about *you*, you know."

"Why, how do you make *that* out? You never heard me play the organ."

"No, doctor, but I've heard you talk about Browning's poetry: and that showed me that you're *metaphysical*, at any rate. So ——"

Extract a Syllogism out of each of the following: and test its correctness :—

96. "Don't talk to me! I've known more rich merchants than you have: and I can tell you not *one* of them was ever an old miser since the world began!"

"And what has that got to do with old Mr. Brown?"

" Why, isn't he very rich ? "

" Yes, of course he is. And what then ? "

" Why, don't you see that it's absurd to call him a miserly merchant? Either he's not a merchant, or he's not a miser!"

97. " It *is* so kind of you to enquire! I'm really feeling a great deal better to-day."

" And is it Nature, or Art, that is to have the credit of this happy change ? "

" Art, I think. The Doctor has given me some of that patent medicine of his."

"Well, I'll never call him a humbug again. There's *somebody*, at any rate, that feels better after taking his medicine !"

98. " No, I don't like you one bit. And I'll go and play with my doll. *Dolls* are never unkind."

" So you like a doll better than a cousin ? Oh you little silly ! "

" Of course I do ! *Cousins* are never kind——at least no cousins *I've* ever seen."

"Well, and what does *that* prove, I'd like to know! If you mean that cousins aren't dolls, who ever said they were ? "

99. " What are you talking about geraniums for ? You can't tell one flower from another, at this distance ! I grant you they're all *red* flowers : it doesn't need a telescope to know *that*."

"Well, some geraniums are red, aren't they?"

"I don't deny it. And what then? I suppose you'll be telling me some of those flowers are geraniums!"

"Of course that's what I should tell you, if you'd the sense to follow an argument! But what's the good of proving anything to *you*, I should like to know?"

100. "Boys, you've passed a fairly good examination, all things considered. Now let me give you a word of advice before I go. Remember that all, who are really anxious to learn, work *hard*."

"I thank you, Sir, in the name of my scholars! And proud am I to think there are *some* of them, at least, that are really *anxious* to learn."

"Very glad to hear it: and how do you make it out to be so?"

"Why, Sir, *I* know how hard they work——some of them, that is. Who should know better?"

Extract from the following speech a series of Syllogisms, or arguments having the form of Syllogisms: and test their correctness.

It is supposed to be spoken by a fond mother, in answer to a friend's cautious suggestion that she is perhaps a *little* overdoing it, in the way of lessons, with her children.

101. "Well, they've got their own way to make in the world. *We* can't leave them a fortune apiece!

And money's not to be had, as *you* know, without money's worth: they must *work* if they want to live. And how are they to work, if they don't know anything? Take my word for it, there's no place for ignorance in *these* times! And all authorities agree that the time to learn is when you're young. One's got no memory afterwards, worth speaking of. A child will learn more in an hour than a grown man in five. So those, that have to learn, must learn when they're young, if ever they're to learn at all. Of course that doesn't do unless children are *healthy*: I quite allow *that*. Well, the doctor tells me no children are healthy unless they've got a good colour in their cheeks. And only just look at my darlings! Why, their cheeks bloom like peonies! Well, now, they tell me that, to keep children in health, you should never give them more than six hours altogether at lessons in the day, and at least two half-holidays in the week. And that's *exactly* our plan, I can assure you! We never go beyond six hours, and every Wednesday and Saturday, as ever is, not one syllable of lessons do they do after their one o'clock dinner! So how you can imagine I'm running any risk in the education of my precious pets is more than *I* can understand, I promise you!"

THE END.

www.ingramcontent.com/pod-product-compliance
Lightning Source LLC
Chambersburg PA
CBHW030410170426
43202CB00010B/1553